Essence of Fractions

Comprehensive Concepts, Transparent Examples,
Practice Exercises, Study Questions

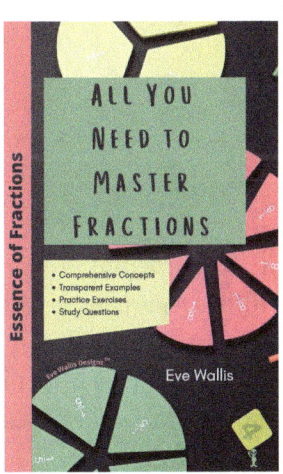

Published by Eve Wallis Designs™

Copyright © 2021 Eve Wallis Designs

HTTPS://EVEWALLIS.WORDPRESS.COM

This material, content, and format are intellectual property of Eve Wallis of "Eve Wallis Designs™". All rights reserved.

No part of this material can be reproduced by any means, or transmitted, or translated into electronic format without the written permission of Eve Wallis.

2021, First Edition

Cover and chapter designs created with Canva.

Copy editing by Eve Wallis and Larry Smith.

Scripture quotations marked NET are from the NET Bible.

ISBN:978-1-7372820-0-6

Contents

Preface	1
1 Introduction to Fractions	**3**
Comparing Fractions	8
Proper, Improper, and Mixed Fractions	14
2 Reducing Fractions	**21**
Reducing with a Greatest Common Divisor	26
Reducing with Prime Factorization	27
3 Multiplying and Dividing Fractions	**33**
Multiplying Fractions	35
Dividing Fractions	42
4 Adding and Subtracting Fractions	**49**
Adding/Subtracting with Common Denominators	51
Adding/Subtracting with Different Denominators	54
5 Fractions with Variables	**63**
Reducing Fractions with Variables	66
Adding/Subtracting Fractions with Variables	68
Multiplying and Dividing Fractions with Variables	72
Appendix	**79**
Checkpoint Answers	81
Study Questions	85
Index	**93**

Preface

This text is designed for adult learners in need of a supplemental resource for learning or reviewing fractions. Minimal background knowledge of mathematics is necessary to make sense of the concepts provided in this book, and the pages are formatted with lists and frames for easy chunking of information. Even if you have not taken a math class in more than 20 years, you can learn and understand the ideas presented in this book by paying close attention to the key concepts and how to use them.

Tucked inside the chapters of this book, you will find explanations of the following key concepts:

- How a fraction is defined.
- How to find equivalent fractions.
- How to convert between mixed numbers and improper fractions.
- How to reduce a fraction to lowest terms.
- How to multiply fractions.
- How to divide fractions.
- How to find a lowest common denominator.
- How to add and subtract fractions with a common denominator.

- How to add and subtract fractions with different denominators.

At the end of each chapter there is a **checkpoint box** containing practice exercises for applying the key concepts corresponding to each chapter. Answers for these checkpoint exercises are provided in the appendix. You will also find **study questions** for reflecting on those key concepts. Look for the ⚷ symbol to find answers for the study questions. Learn these key concepts to understand the essence of fractions.

Introduction to Fractions

> **🔑 Key Concepts**
>
> - How a fraction is defined.
>
> - How to find equivalent fractions.
>
> - How to convert between mixed numbers and improper fractions.

Definition of a Fraction

There is a pie in the fridge that is cut into 8 equal slices. After one day, only 4 of the slices are remaining.

This means there are 4 out of 8 slices remaining or

$$\frac{4}{8} \text{ slices remaining}$$

This number is called a fraction. *Fractions give us a way to represent numbers that are some part of a whole amount.* (🔑) In this fraction of pie servings, the top number, 4, represents the part amount (the remaining part of the pie) and the bottom number, 8, represents the whole amount (the whole pie).

CHAPTER 1. INTRODUCTION TO FRACTIONS

Then 4 slices out of 8 slices is also half of the pie,

4 out of 8 or $\frac{1}{2}$ of the pie

and 2 slices out of 8 slices would be a quarter of the pie,

2 out of 8 or $\frac{1}{4}$ of the pie

and 6 slices out of 8 slices would be three quarters of the pie.

6 out of 8 or $\frac{3}{4}$ of the pie

We can also represent the whole pie as a fraction.

$\frac{8}{8}$ or 1 whole pie

In general, the *part* or top number in a fraction is called the *numerator* of the fraction and the *whole* or bottom number in a fraction is called the *denominator* of the fraction.

> ⚙ **Fraction Definition**
>
> A fraction is a number of the form
>
> $$\frac{n}{d}$$
>
> where the top number n is called the **numerator** and the bottom number d is called the **denominator**.
>
>

Whole numbers [1] can be fractions by placing the whole number over 1.

$$\frac{0}{1}, \frac{1}{1}, \frac{2}{1}, \frac{3}{1}, \frac{4}{1}, \ldots$$

Integers [2] can be fractions by placing the integer over 1.

$$\ldots \frac{-3}{1}, \frac{-2}{1}, \frac{-1}{1}, \frac{0}{1}, \frac{1}{1}, \frac{2}{1}, \frac{3}{1}, \ldots$$

Fractions with denominator zero are undefined.

For example, the fraction $\frac{3}{0}$ represents 3 parts of a 0 whole, which does not make any sense.

Any number that can be represented in the form $\frac{n}{d}$ is part of the fraction number set.

Fractions are very useful. We can demonstrate the division of

---

[1] Counting numbers and zero {0,1,2,3,4, ...}
[2] Zero, counting numbers, and their opposites {... -3, -2, -1, 0,1,2,3, ...}

CHAPTER 1. INTRODUCTION TO FRACTIONS

many different things with fractions. When we peel and slice an apple we are dividing the apple into fractional parts of the whole apple. When we buy a pizza and share it with our friends, we divide it into slices and serve each friend a fraction of the whole pizza. We also use fractions in baking, as many recipes call for a fractional amount of different ingredients. We also use fractions with measurements and money. If we want to buy half a pound of grapes that are priced per pound, then we will need to find half the cost per pound to determine what we will pay for the grapes. Fractions are used in many different applications, and it is important to learn how to use fractions for making decisions every day.

Comparing Fractions

In this section, we will learn how to find equivalent fractions and how to compare fractions that are not equivalent.

Equivalent Fractions

Two or more fractions are **equivalent fractions** if they represent the same numerical value.

We saw equivalent fractions earlier with the pie example.

$\frac{4}{8}$ or $\frac{1}{2}$ of the pie

The fractions in this example, $\frac{4}{8}$ and $\frac{1}{2}$, are equivalent because they both represent half of the pie.

Fractions where the part amount is half of the whole amount, such as $\frac{2}{4}, \frac{3}{6}, \frac{4}{8}$ etc., are all equivalent to the fraction $\frac{1}{2}$.

Fractions, where the part amount is the same as the whole amount, such as $\frac{2}{2}, \frac{3}{3}, \frac{4}{4}$, etc., are all equivalent to 1.

We can find other equivalent fractions using multiplication or division.

> ## ⚙ Finding Equivalent Fractions
>
> There are two ways to find an equivalent fraction:
>
> 1. Multiply the numerator and the denominator by the same number.
>
> $$\frac{1}{2} = \frac{1 \times 4}{2 \times 4} = \frac{4}{8}$$
>
> 2. Divide the numerator and the denominator by the same number.
>
> $$\frac{4}{8} = \frac{4 \div 4}{8 \div 4} = \frac{1}{2}$$

CHAPTER 1. INTRODUCTION TO FRACTIONS

Example 1.1 Find equivalent fractions with denominator 9.

1. $\frac{2}{3}$
2. $\frac{18}{27}$

1. *Solution:* To get a denominator of 9, we would have to multiply the denominator by 3. Therefore, to get an equivalent fraction with denominator 9 we will *multiply the numerator and the denominator by 3*.

$$\frac{2}{3} = \frac{2 \times 3}{3 \times 3} = \frac{6}{9}$$

2. *Solution:* To get a denominator of 9, we would have to divide the denominator by 3. Therefore, to get an equivalent fraction with denominator 9 we will *divide the numerator and the denominator by 3*.

$$\frac{18}{27} = \frac{18 \div 3}{27 \div 3} = \frac{6}{9}$$

Fractions that are not Equivalent

We may **compare fractions that are not equivalent** with the following symbols:

- **Less Than** Use the "<" symbol to denote that a number is "less than" another number. (e.g. $\frac{1}{3} < \frac{2}{3}, \frac{1}{4} < \frac{3}{4}$)

- **Greater Than** Use the ">" symbol to denote that a number is "greater than" another number. (e.g. $\frac{2}{3} > \frac{1}{3}, \frac{3}{4} > \frac{1}{4}$)

We may compare some fractions without computation if they have the same numerator, the same denominator, or they can be compared to a $\frac{1}{2}$ or 1.

- **Same Denominators** If two fractions have the same denominator, the fraction with the larger numerator is *greater than* the fraction with the smaller numerator.

  Example:

  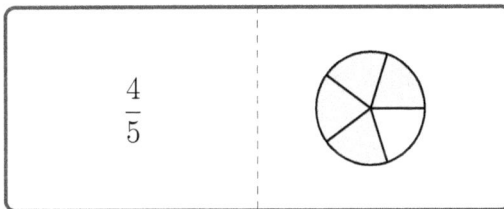

  is greater than

  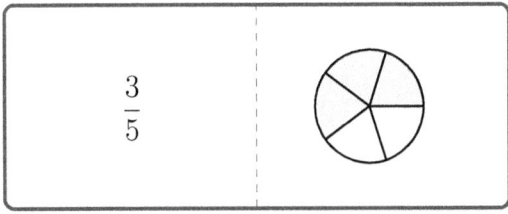

  Furthermore, $\frac{9}{10} > \frac{8}{10}, \frac{3}{4} > \frac{1}{4}, etc.$

- **Same Numerators** If two fractions have the same numerator, the fraction with a larger denominator is *smaller than* the fraction with a smaller denominator.

  Example:

CHAPTER 1. INTRODUCTION TO FRACTIONS

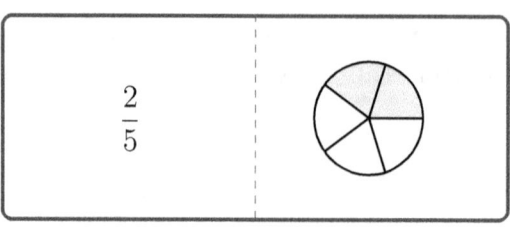

is smaller than

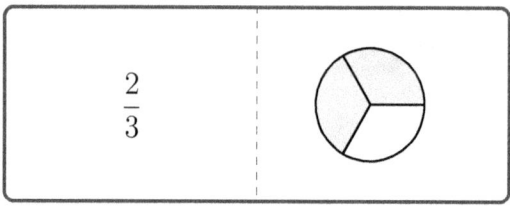

Furthermore, $\frac{1}{100} < \frac{1}{10}, \frac{3}{4} < \frac{3}{2}, etc.$

- **Compared to a Half** A fraction that is less than a $\frac{1}{2}$ is less than a fraction that is greater than a $\frac{1}{2}$.
(e.g. $\frac{1}{4} < \frac{2}{4} = \frac{1}{2}$ and $\frac{5}{6} > \frac{3}{6} = \frac{1}{2}$, therefore, $\frac{1}{4} < \frac{5}{6}$)

- **Compared to a Whole** A fraction that is less than 1 whole is less than a fraction that is greater than 1 whole.
(e.g. $\frac{2}{3} < \frac{3}{3} = 1$ and $\frac{3}{2} > \frac{3}{3} = 1$, therefore, $\frac{2}{3} < \frac{3}{2}$)

Some fractions cannot be compared with the methods given above. For example, if we want to know how the fraction $\frac{2}{3}$ compares to the fraction $\frac{3}{4}$, how do we do it? They have different numerators, different denominators, they are both greater than a half and less than 1, so how do we determine which fraction is greater?

To determine which is greater, we must change $\frac{2}{3}$ and $\frac{3}{4}$ to equivalent fractions with the same denominator.

$$\frac{2}{3} = \frac{2 \times 4}{3 \times 4} = \frac{8}{12}$$

$$\frac{3}{4} = \frac{3 \times 3}{4 \times 3} = \frac{9}{12}$$

This makes it easier to see that $\frac{2}{3}$ is less than $\frac{3}{4}$.

$$\frac{2}{3} = \frac{8}{12} < \frac{9}{12} = \frac{3}{4}$$

In conclusion, any two fractions that do not have the same denominator may be compared by changing them to equivalent fractions that *do* have the same denominators.

✿ Comparing Fractions with Different Denominators

If two fractions have different denominators:

- Change the fractions to equivalent fractions with the same denominator.

- The fraction with the larger numerator is the greater fraction.

Example 1.2 Which is greater? $\frac{23}{25}$ or $\frac{41}{50}$

- Change to equivalent fractions with the same denominator.

    Multiply the 1st fraction by 2 to get the same denominator as the 2nd fraction.

    $$\frac{23 \times 2}{25 \times 2} = \frac{46}{50}$$

- The fraction with the larger numerator is the larger fraction.

CHAPTER 1. INTRODUCTION TO FRACTIONS

$$\frac{23}{25} = \frac{46}{50} > \frac{41}{50}$$

Proper, Improper, and Mixed Fractions

There are three general types of fractions that we will use in this text: proper fractions, improper fractions, and mixed fractions.

- **Proper fractions** are fractions that are less than 1.

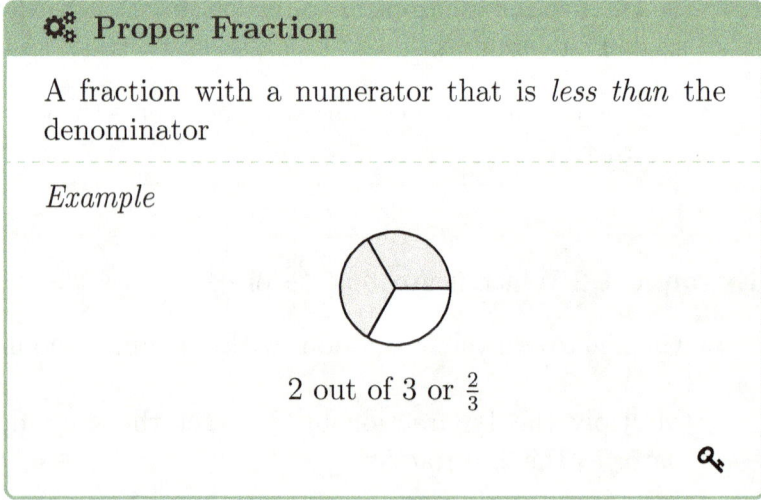

> **Proper Fraction**
>
> A fraction with a numerator that is *less than* the denominator
>
> *Example*
>
> 2 out of 3 or $\frac{2}{3}$

- **Improper fractions** are fractions that are greater than or equal to 1.

> ## ⚙ Improper Fractions
>
> A fraction with a numerator that is *more than* the denominator.
>
> *Example*
>
>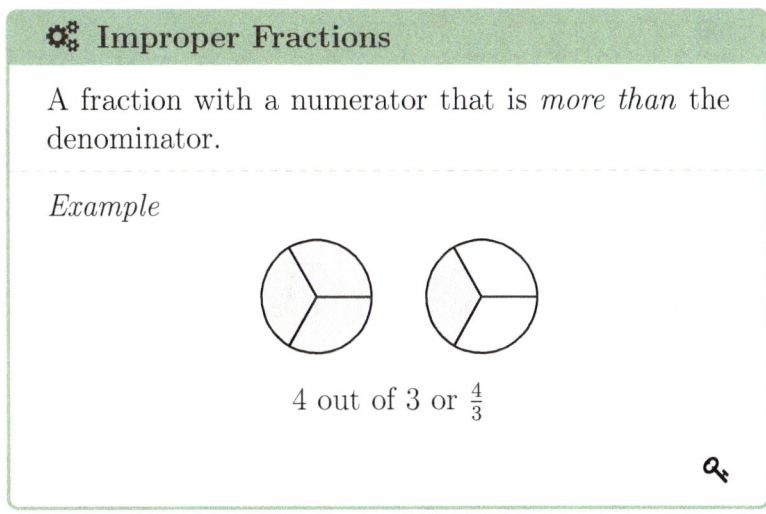
>
> 4 out of 3 or $\frac{4}{3}$

- **Mixed fractions** are mixed numbers that consist of a whole number and a proper fraction.

> ## ⚙ Mixed Number
>
> A number that consists of a whole number and a proper fraction.
>
> *Example*
>
>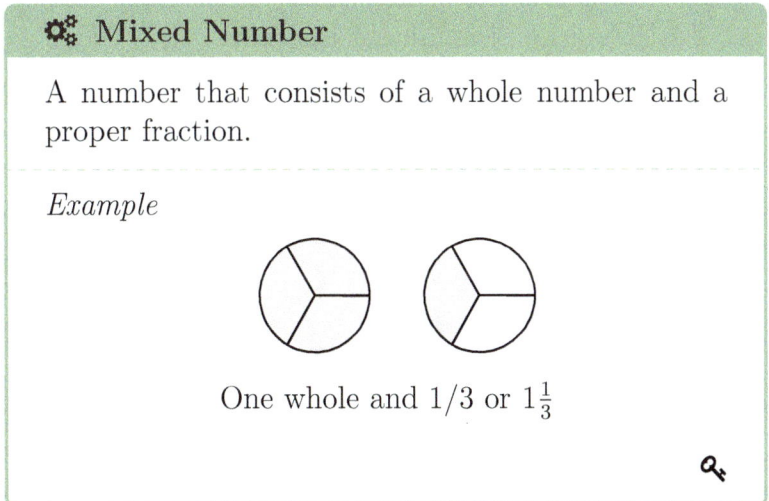
>
> One whole and 1/3 or $1\frac{1}{3}$

Notice that the improper fraction $\frac{4}{3}$ and the mixed number $1\frac{1}{3}$ represent the same amount. Mixed numbers and improper fractions both represent a fraction amount that is greater than or

CHAPTER 1. INTRODUCTION TO FRACTIONS

equal to 1.

It is also possible to convert a mixed number to an improper fraction and an improper fraction to a mixed number.

> **⚙ Improper Fraction ⊙ Mixed Number**
>
> To convert an improper fraction to a mixed number:
>
> - Divide the numerator by the denominator.
>
> - Make the quotient the whole number part of the mixed number.
>
> - Make the remainder the numerator part of the mixed number.
>
> - Keep the denominator the same.

Example 1.3 Convert $\frac{4}{3}$ to a mixed number.

- Divide the numerator by the denominator.

$$3\overline{)4} \\ \underline{3} \\ 1$$ with quotient 1

- Make the quotient the whole number part of the mixed number. $1\frac{\square}{\square}$

- Make the remainder the numerator part of the mixed number. $1\frac{1}{\square}$

- Keep the denominator the same $1\frac{1}{3}$

Therefore,

$$\frac{4}{3} = \quad 3\overline{)4}^{\,1}\underline{}_{3}_{1} = 1\frac{1}{3}$$

> 📝 **In Plain English**
>
> 3 goes into 4, 1 time, with 1 left over.

⚙ Mixed Number ➡ Improper Fraction

To convert a mixed number to an improper fraction

- Multiply the denominator and the whole number part

- Add the product to the numerator and make this the numerator part of the improper fraction.

- Keep the denominator the same.

CHAPTER 1. INTRODUCTION TO FRACTIONS

Example 1.4 Convert $1\frac{1}{3}$ to an improper fraction.

- Multiply the denominator and the whole number part. $3 \times 1 = 3$
- Add the product to the numerator $3 + 1 = 4$
- And make this the numerator part of the improper fraction. $\frac{4}{\square}$
- Keep the denominator the same. $\frac{4}{3}$

Therefore,
$$1\frac{1}{3} = \frac{(3 \times 1) + 1}{3} = \frac{4}{3}$$

☑ Checkpoint - Introduction to Fractions

1. Fill in each blank with the correct symbol to compare the fractions: $<, >,$ or $=$.

   a) $\frac{4}{9}$ ___ $\frac{7}{9}$ d) $\frac{5}{10}$ ___ $\frac{6}{12}$

   b) $\frac{1}{9}$ ___ $\frac{1}{7}$ e) $\frac{7}{2}$ ___ $\frac{2}{7}$

   c) $\frac{3}{8}$ ___ $\frac{4}{5}$ f) $\frac{117}{144}$ ___ $\frac{147}{144}$

2. Convert each mixed number to an improper fraction and each improper fraction to a mixed number.

   a) $5\frac{3}{7}$ c) $\frac{22}{7}$

   b) $3\frac{3}{4}$ d) $\frac{42}{5}$

3. Change each fraction to an equivalent fraction with denominator 25

   a) $\frac{1}{5}$

   b) $\frac{2}{50}$

   c) $\frac{48}{100}$

   d) $\frac{65}{125}$

4. Consider the circle given below.

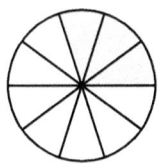

   a) What fraction of the circle is shaded?

   b) What fraction of the circle is not shaded?

   (Answers in the Appendix.)

Reducing Fractions

> 🔑 **Key Concepts**
>
> - How to reduce a fraction.
>
> - How to reduce a fraction to lowest terms with a greatest common divisor.
>
> - How to reduce a fraction to lowest terms with prime factorization.

Introduction to Reducing Fractions

We can reduce a fraction to a simpler fraction, if both parts of the fraction share a **common divisor**, a number that divides evenly into the the numerator and the denominator.

For example, the fraction $\frac{4}{8}$ can be reduced to $\frac{1}{2}$ by dividing 4 and 8 by *common divisor* 4.

$$\frac{4 \div 4}{8 \div 4} = \frac{1}{2}$$

> ⚙ **Reducing Fractions**
>
> To reduce a fraction, divide both parts by a common divisor.

CHAPTER 2. REDUCING FRACTIONS

Example 2.1 Reduce the fraction $\frac{3}{6}$

Solution:
Divide both parts of the fraction by common divisor 3 to reduce the fraction.

$$\frac{3 \div 3}{6 \div 3} = \frac{1}{2}$$

Sometimes it can be tricky to find common divisors if a fraction contains large numbers. The following divisibility rules are helpful for finding divisors of large numbers.

Divisibility Rules:

- **Divisible by 2** A number is divisible by 2 (has 2 as a divisor) if it ends in an even number. (e.g. 22, 36, 58, etc.)

- **Divisible by 3** A number is divisible by 3 (has 3 as a divisor) if the sum of the digits is divisible by 3. (e.g. 18 is divisible by 3 because $1 + 8 = 9$ which is divisible by 3.)

- **Divisible by 4** A number is divisible by 4 if the last two digits are divisible by 4. (e.g. 328 is divisible by 4 because 28 is divisible by 4.)

- **Divisible by 5** A number is divisible by 5 if it ends in a 5 or a 0. (e.g. 5, 10, 15, 20, 25, etc.)

- **Divisible by 6** A number is divisible by 6 if it is divisible by 2 and 3. (e.g. 18 is divisible by 6 because its ends in an even number and $1 + 8 = 9$ is divisible by 3.)

- **Divisible by 9** A number is divisible by 9 if the sum of its digits is divisible by 9. (e.g. 36 is divisible by 9 because $3 + 6 = 9$ which is divisible by 9.)

- **Divisible by 10** A number is divisible by 10 if it ends in a 0. (e.g. 10, 20, 30, 40, etc.)

Example 2.2 Reduce the fraction.

$$\frac{27}{54}$$

Solution:
3 is a divisor of 27, since $2 + 7 = 9$ and 9 is divisible by 3.
3 is also a divisor of 54, since $5 + 4 = 9$ and 9 is divisible by 3.

Since 3 is a common divisor of 54 and 27, we can divide both parts of the fraction by 3 to reduce the fraction.

$$\frac{27 \div 3}{54 \div 3} = \frac{9}{18}$$

Notice that the fraction 9/18 can still be reduced, since 9 is also a common divisor of 9 and 18. Dividing by 9 again would lead to the fraction $\frac{1}{2}$, which cannot be reduced any further.

$$\frac{9 \div 9}{18 \div 9} = \frac{1}{2}$$

A fraction that cannot be reduced any further is a fraction that is *reduced to lowest terms*.

CHAPTER 2. REDUCING FRACTIONS

Reducing with a Greatest Common Divisor

We can reduce a fraction to lowest terms, if we divide each part of the fraction by the **greatest common divisor**, the largest natural number that divides evenly into both parts of the fraction. (🔑)

For example, 27 and 54 have common divisors 1, 3, 9, and 27, but 27 is the greatest common divisor. If we divide $\frac{27}{54}$ by this greatest common divisor, the immediate result will be a fraction that is reduced to lowest terms.

$$\frac{27 \div 27}{54 \div 27} = \frac{1}{2}$$

> **⚙ Reducing Fractions - Greatest Common Divisor**
>
> To reduce fractions with a greatest common divisor:
>
> - Find the greatest common divisor of the numerator and the denominator.
>
> - Divide the numerator and the denominator by the greatest common divisor to reduce to lowest terms.
>
> 🔑

> **✎ In Plain English**
>
> To reduce a fraction to lowest terms, find the largest number that divides evenly into each part, and divide each part by that number.

Example 2.3 Reduce each fraction to lowest terms.

1. $\frac{12}{18}$
2. $\frac{18}{27}$

1. *Solution:* Find the greatest common divisor of 12 and 18.

   12 has divisors 1, 2, 3, 4, 6, and 12.
   18 has divisors 1, 2, 3, 6, 9, and 18.
   The greatest common divisor is 6.

   Divide 12 and 18 by 6 to reduce to lowest terms.

   $$\frac{12}{18} = \frac{12 \div 6}{18 \div 6} = \frac{2}{3}$$

2. *Solution:* Find the greatest common divisor of 18 and 27.

   18 has divisors 1, 2, 3, 6, 9, and 18.
   27 has divisors 1, 3, 9, and 27.
   The greatest common divisor is 9.

   Divide 18 and 27 by 9 to reduce to lowest terms.

   $$\frac{18}{27} = \frac{18 \div 9}{27 \div 9} = \frac{2}{3}$$

Reducing with Prime Factorization

We can also reduce a fraction to lowest terms if we write each part as a product of its prime factors.

CHAPTER 2. REDUCING FRACTIONS

Factors are numbers we multiply together to get a larger number. (e.g. 1, 2, 3, and 6 are factors of 6 because $1 \times 6 = 6$ and $2 \times 3 = 6$) (🔎)

Numbers that have more than two factors are called **composite numbers** . (e.g. 4, 6, 8, 9, etc.)

Numbers that have only two factors, 1 and that number, are called **prime numbers** . (e.g. 2, 3, 5, 7, etc.) (🔎)

A **prime factor** is a factor that is also a prime number. (e.g. 1, 2, 3, and 6 are factors of 6 but only 2 and 3 are prime factors of 6) (🔎)

Finding all the prime factors of a number is called **prime factorization**.

> ⚙ **Prime Factorization**
>
> To find the prime factorization of a number:
>
> - Divide the number by its lowest prime factor.
>
> - Divide the quotient by its lowest prime factor.
>
> - Repeat until there are no more prime factors.
>
> - Write the number as a product of its prime factors.

Example 2.4 Write 28 as a product of its prime factors.

- Divide 28 by its lowest prime factor 2. $28 \div 2 = 14$
- Divide 14 by its lowest prime factor 2. $14 \div 2 = 7$
- 7 is already a prime factor.
- The prime factors of 28 are 2, 2, and 7 such that the prime factorization is $2 \cdot 2 \cdot 7$.

When finding prime factors, it may be helpful to make a table.

Draw an L. Then place the original number, 28, on the inside and the lowest prime factor of 28 on the outside.

$$\begin{array}{c|c} 2 & 28 \\ \hline \end{array}$$

Then divide 28 by 2 and place the result below 28.

$$\begin{array}{c|c} 2 & 28 \\ \hline & 14 \end{array}$$

Draw another line under 14 and place the lowest prime factor of 14 on the outside.

$$\begin{array}{c|c} 2 & 28 \\ \hline 2 & 14 \\ \hline \end{array}$$

Then divide 14 by 2 and place the result below 14.

$$\begin{array}{c|c} 2 & 28 \\ \hline 2 & 14 \\ \hline & 7 \end{array}$$

CHAPTER 2. REDUCING FRACTIONS

This table makes it easier to see that the prime factors are 2, 2, and 7, because they are the numbers on the outside and at the bottom of the table.

If we can find the primes factors of each part of a fraction, then we can reduce the fraction to lowest terms by canceling the common factors between those fraction parts. Sometimes this method is easier than finding a greatest common divisor, particularly if the fraction contains large numbers.

> **Reducing Fractions - Prime Factorization**
>
> To reduce fractions with prime factorization.
>
> - Find the prime factorization of the numerator and the denominator (writing each as a product of its prime factors)
>
> - Cancel all common factors between the numerator and the denominator.

Example 2.5 Reduce each fraction with prime factorization.

1. $\frac{12}{18}$
2. $\frac{18}{27}$

1. *Solution:* Find the prime factors of 12 and 18.

$$
\begin{array}{r|r}
2 & 12 \\
2 & 6 \\ \hline
 & 3
\end{array}
\qquad\qquad
\begin{array}{r|r}
2 & 18 \\
3 & 9 \\ \hline
 & 3
\end{array}
$$

Write each part of the fraction as a product of its prime factors.

$$\frac{12}{18} = \frac{2 \cdot 2 \cdot 3}{2 \cdot 3 \cdot 3}$$

Cancel common factors 2 and 3.

$$\frac{\cancel{2} \cdot 2 \cdot \cancel{3}}{\cancel{2} \cdot 3 \cdot \cancel{3}} = \frac{2}{3}$$

2. *Solution:* Find the prime factors of 18 and 27.

$$
\begin{array}{r|r}
3 & 27 \\
3 & 9 \\ \hline
 & 3
\end{array}
\qquad\qquad
\begin{array}{r|r}
2 & 18 \\
3 & 9 \\ \hline
 & 3
\end{array}
$$

Write each part of the fraction as a product of its prime factors.

$$\frac{18}{27} = \frac{2 \cdot 3 \cdot 3}{3 \cdot 3 \cdot 3}$$

Cancel common factors 3 and 3.

$$\frac{2 \cdot \cancel{3} \cdot \cancel{3}}{3 \cdot \cancel{3} \cdot \cancel{3}} = \frac{2}{3}$$

CHAPTER 2. REDUCING FRACTIONS

> ☑ **Checkpoint - Reducing Fractions**
>
> 1. Find the greatest common divisor for each group of numbers.
>
>   a) 24 and 32
>
>   b) 45 and 95
>
>   c) 12, 16, and 18
>
>   d) 28, 42, and 84
>
> 2. Find the prime factorization of each number.
>
>   a) 32
>
>   b) 84
>
>   c) 135
>
>   d) 625
>
> 3. Reduce each fraction to lowest terms.
>
>   a) $\frac{9}{27}$
>
>   b) $\frac{24}{42}$
>
>   c) $\frac{135}{625}$
>
>   d) $\frac{32}{84}$
>
> (Answers in the Appendix.)

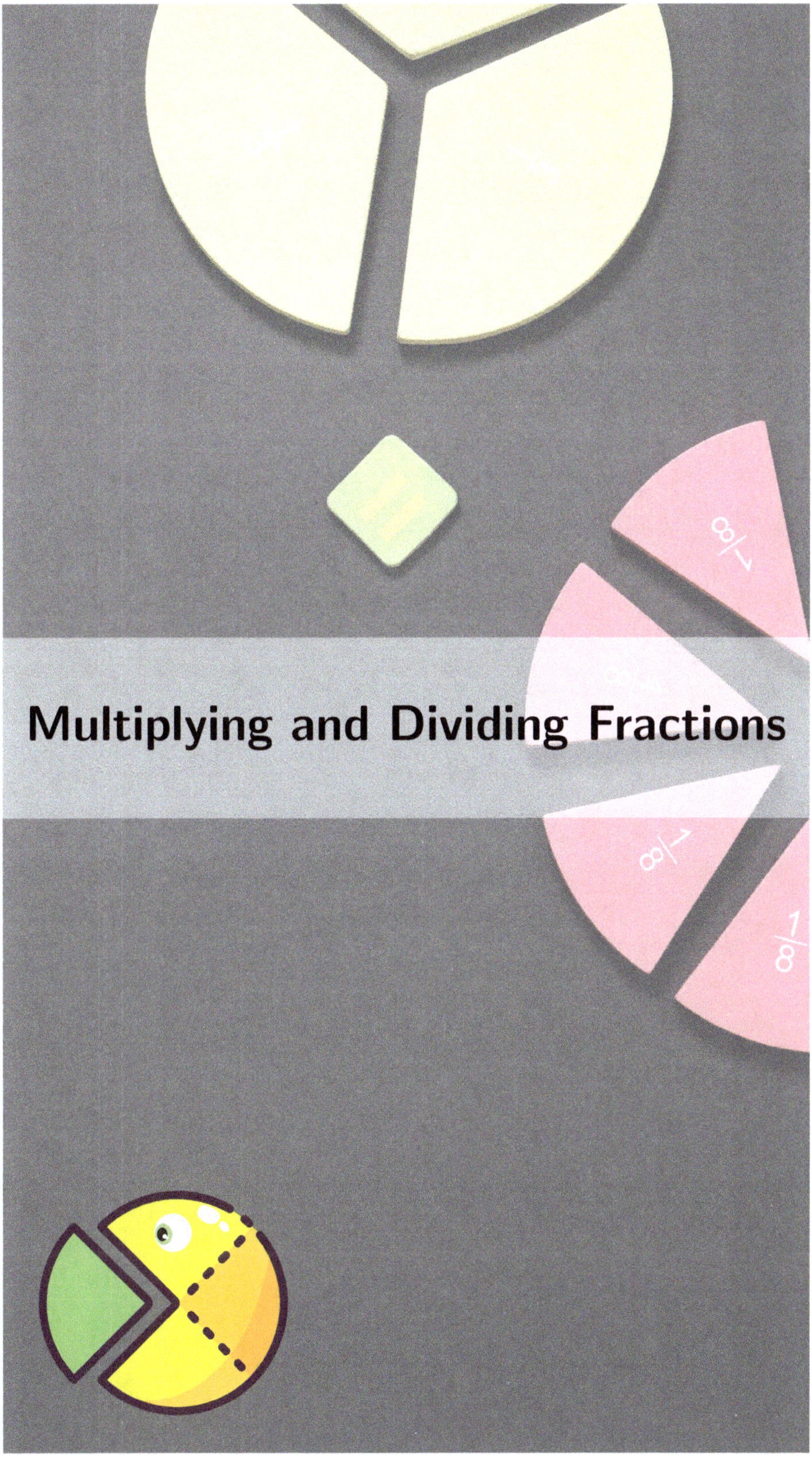

Multiplying and Dividing Fractions

> 🔑 **Key Concepts**
> - How to multiply fractions.
> - How to divide fractions.

Multiplying Fractions

When we multiply two proper fractions we get a smaller amount as the result.

For example, $\frac{2}{3}$ represents "2 parts of 3",

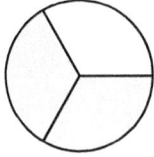

and half of $\frac{2}{3}$ represents "1 part of 3"

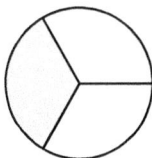

In conclusion, $\frac{1}{2} \times \frac{2}{3} = \frac{1}{3}$

In general, we may use the following steps for multiplying fractions:

CHAPTER 3. MULTIPLYING AND DIVIDING FRACTIONS

⚙ Multiplying Fractions

When multiplying fractions

- Multiply across (Multiply the numerators; Multiply the denominators)

$$\frac{1}{2} \cdot \frac{2}{3} = \frac{1 \cdot 2}{2 \cdot 3} = \frac{2}{6}$$

- Reduce to lowest terms

$$\frac{2}{6} = \frac{2 \div 2}{6 \div 2} = \frac{1}{3}$$

The notation we use for multiplication may vary between a times symbol, parentheses, and a dot.

Times	Parentheses	dot
3×8	$(3)(8)$	$3 \cdot 8$

Example 3.1 Multiply and reduce each answer to lowest terms.

1. $\frac{3}{4} \cdot \frac{2}{9}$
2. $\frac{11}{12} \cdot 4$

1. *Solution:* Multiply across

$$\frac{3}{4} \cdot \frac{2}{9} = \frac{3 \cdot 2}{4 \cdot 9} = \frac{6}{36}$$

Reduce to lowest terms with greatest common divisor 6.

$$\frac{6 \div 6}{36 \div 6} = \frac{1}{6}$$

2. *Solution:* Change the second factor to a fraction by placing 4 over 1.
$$\frac{11}{12} \cdot 4 = \frac{11}{12} \cdot \frac{4}{1}$$
Multiply across.
$$\frac{11}{12} \cdot \frac{4}{1} = \frac{11 \cdot 4}{12 \cdot 1} = \frac{44}{12}$$
Reduce to lowest terms with greatest common divisor 4.
$$\frac{44 \div 4}{12 \div 4} = \frac{11}{3}$$
Convert to a mixed number by dividing.
$$\frac{11}{3} = 3\overline{)11} = 3\frac{2}{3}$$

> ✎ In Plain English
>
> 3 goes into 11, 3 times, with 2 left over.

If we are **multiplying mixed numbers** then we will need to convert them to improper fractions before applying the rules for multiplying fractions.

We can see this with the following example.

CHAPTER 3. MULTIPLYING AND DIVIDING FRACTIONS

Example 3.2 Multiply and reduce each answer to lowest terms.

1. $1\frac{3}{4} \cdot 5\frac{1}{9}$
2. $2\frac{11}{12} \cdot 4$

1. *Solution:* First convert each mixed number to an improper fraction

$$1\frac{3}{4} = \frac{4 \cdot 1 + 3}{4} = \frac{7}{4}$$

$$5\frac{1}{9} = \frac{9 \cdot 5 + 1}{9} = \frac{46}{9}$$

Multiply across

$$\frac{7}{4} \cdot \frac{46}{9} = \frac{7 \cdot 46}{4 \cdot 9} = \frac{322}{36}$$

Reduce with prime factorization.

$$\frac{322}{36} = \frac{\cancel{2} \cdot 161}{\cancel{2} \cdot 2 \cdot 3 \cdot 3} = \frac{161}{18}$$

Then convert back to a mixed number

$$\frac{161}{18} = 18 \overline{)161} = 8\frac{17}{18}$$

with long division showing 8, 144, remainder 17.

> 📝 In Plain English
>
> 18 goes into 161, 8 times, with 17 left over.

2. *Solution:* Convert each factor to an improper fraction.

$$2\frac{11}{12} \cdot 4 = \frac{35}{12} \cdot \frac{4}{1}$$

Multiply across

$$\frac{35 \cdot 4}{12 \cdot 1} = \frac{140}{12}$$

Reduce to lowest terms with the greatest common divisor.

$$\frac{140 \div 4}{12 \div 4} = \frac{35}{3}$$

It is also possible to reduce before multiplying fractions, by factoring each part and canceling common factors.

$$\frac{35}{12} \cdot \frac{4}{1} = \frac{5 \cdot 7}{3 \cdot \cancel{4}} \cdot \frac{\cancel{4}}{1} = \frac{35}{3}$$

Canceling before multiplying is tricky, so we will consider another example.

CHAPTER 3. MULTIPLYING AND DIVIDING FRACTIONS

Example 3.3 Cancel common factors then multiply

$$\frac{35}{12} \cdot \frac{28}{125}$$

Factor each part

$$\frac{35}{12} \cdot \frac{28}{125} = \frac{5 \cdot 7}{3 \cdot 4} \cdot \frac{4 \cdot 7}{5 \cdot 25}$$

Cancel Common Factors

$$\frac{\cancel{5} \cdot 7}{3 \cdot \cancel{4}} \cdot \frac{\cancel{4} \cdot 7}{\cancel{5} \cdot 25} = \frac{7}{3} \cdot \frac{7}{25}$$

Multiply across

$$\frac{7 \cdot 7}{3 \cdot 25} = \frac{49}{75}$$

Notice that we do not need to reduce at the end because we already reduced when we canceled common factors.

> ⚠ **CAUTION**
>
> We may only cancel between a numerator and a denominator.
>
> $$\frac{7}{3} \cdot \frac{7}{25} = \frac{\cancel{7}}{3} \cdot \frac{7}{\cancel{25}} = \frac{1}{75}$$
>
> Canceling between two numerators or two denominators is not permitted.

When we are **multiplying more than two fractions** we can still apply the same rules, and multiply across.

Example 3.4 Multiply and reduce to lowest terms.
$$\frac{1}{7} \cdot \frac{3}{4} \cdot \frac{14}{15}$$

Solution:

Multiply across
$$\frac{1 \cdot 3 \cdot 14}{7 \cdot 4 \cdot 15} = \frac{42}{420}$$

Reduce to lowest terms

$$\frac{42}{420} = \frac{\not{2} \cdot \not{3} \cdot \not{7}}{\not{2} \cdot 2 \cdot \not{3} \cdot 5 \cdot \not{7}} = \frac{1}{10}$$

Example 3.5 You need to buy $3\frac{1}{2}$ lbs of grapes that cost $6/lb. How much will you pay for the grapes?

Solution: Multiply $3\frac{1}{2} \cdot 6$

Convert to proper fractions.

$$3\frac{1}{2} \cdot 6 = \frac{7}{2} \cdot \frac{6}{1}$$

Multiply across.

$$\frac{7 \cdot 6}{2 \cdot 1} = \frac{42}{2}$$

Reduce to lowest terms.

CHAPTER 3. MULTIPLYING AND DIVIDING FRACTIONS

$$\frac{42 \div 2}{2 \div 2} = \frac{21}{1} = 21$$

You will pay $21 for the grapes.

Dividing Fractions

Division is a process of dividing a whole into separate parts.

Dividing 1 by $\frac{1}{3}$ is dividing 1 whole into 3 parts that are each $\frac{1}{3}$ of a whole.

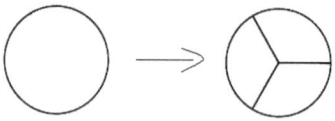

$$1 \div \frac{1}{3} = 3$$

Dividing 2 by $\frac{1}{3}$ is dividing 2 wholes into 6 parts that are each $\frac{1}{3}$ of a whole.

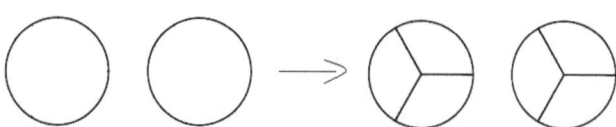

$$2 \div \frac{1}{3} = 6$$

Notice that when we are dividing by $\frac{1}{3}$ we are actually multiplying by 3.

$$1 \div \frac{1}{3} = 1 \cdot \frac{3}{1} = 3$$

$$2 \div \frac{1}{3} = 2 \cdot \frac{3}{1} = 6$$

When we divide by a fraction we actually multiply by the reciprocal of that fraction. The **reciprocal** of a fraction is that fraction *flipped* over.

The reciprocal of $\frac{1}{3}$ is $\frac{3}{1}$.

The reciprocal of $\frac{2}{3}$ is $\frac{3}{2}$.

etc.

We may use the following steps when dividing fractions.

⚙ Dividing Fractions

When dividing fractions

- Change to multiplication of the 1st fraction times the reciprocal of the 2nd fraction.

$$\frac{1}{3} \div \frac{2}{3} = \frac{1}{3} \cdot \frac{3}{2}$$

CHAPTER 3. MULTIPLYING AND DIVIDING FRACTIONS

- Multiply across
$$\frac{1 \cdot 3}{3 \cdot 2} = \frac{3}{6}$$

- Reduce to lowest terms.
$$\frac{3 \div 3}{6 \div 3} = \frac{1}{2}$$

✏ In Plain English

Flip the 2nd fraction and multiply.

The notation we use for division may vary between a division symbol, a fraction bar, and a division bar.

Symbol	Fraction Bar	Division Bar
$8 \div 4$	$\frac{8}{4}$	$4\overline{)8}$

Example 3.6 Divide and reduce each answer to lowest terms.

1. $\frac{3}{4} \div \frac{9}{12}$
2. $\frac{5}{13} \div 5$

1. *Solution:* Flip the 2nd fraction and change to multiplication.
$$\frac{3}{4} \cdot \frac{12}{9}$$

44

Multiply across.
$$\frac{3 \cdot 12}{4 \cdot 9} = \frac{36}{36}$$

Reduce to lowest terms.
$$\frac{36 \div 36}{36 \div 36} = \frac{1}{1} = 1$$

> **Math Bit**
>
> It is also possible to cancel common factors, before multiplying, to reduce.
> $$\frac{3}{4} \cdot \frac{12}{9} = \frac{\cancel{3}}{\cancel{4}} \cdot \frac{\cancel{3} \cdot \cancel{4}}{\cancel{3} \cdot \cancel{3}} = \frac{1}{1} \cdot \frac{1}{1} = 1$$

2. *Solution:* Change the 5 to a fraction by placing it over 1.
$$\frac{5}{13} \div 5 = \frac{5}{13} \div \frac{5}{1}$$

Flip the 2nd fraction and change to multiplication.
$$\frac{5}{13} \div \frac{5}{1} = \frac{5}{13} \cdot \frac{1}{5}$$

Cancel common factors and multiply across.
$$\frac{\cancel{5}}{13} \cdot \frac{1}{\cancel{5}} = \frac{1 \cdot 1}{13 \cdot 1} = \frac{1}{13}$$

CHAPTER 3. MULTIPLYING AND DIVIDING FRACTIONS

When **dividing mixed numbers**, we have to first convert the mixed numbers to improper fractions. Then divide with the rules for dividing fractions.

Example 3.7 Divide and reduce to lowest terms.

$$5\frac{1}{2} \div 2\frac{3}{4}$$

Solution: Convert the mixed numbers to improper fractions.

$$5\frac{1}{2} \div 2\frac{3}{4} = \frac{11}{2} \div \frac{11}{4}$$

Flip the 2nd fraction and change to multiplication.

$$\frac{11}{2} \div \frac{11}{4} = \frac{11}{2} \cdot \frac{4}{11}$$

Cancel common factors and multiply across.

$$\frac{\cancel{11}}{\cancel{2}} \cdot \frac{2 \cdot 2}{\cancel{11}} = \frac{1}{1} \cdot \frac{2}{1} = 2$$

When **dividing more than two fractions** we can divide from left to right, beginning with the first two fractions on the left.

Example 3.8 Divide and reduce to lowest terms.

$$\frac{5}{6} \div \frac{5}{7} \div \frac{11}{6}$$

Solution: Divide the first two fractions.

$$\frac{5}{6} \div \frac{5}{7} = \frac{5}{6} \cdot \frac{7}{5} = \frac{7}{6}$$

Divide the result and the third fraction.

$$\frac{7}{6} \div \frac{11}{6} = \frac{7}{6} \cdot \frac{6}{11} = \frac{7}{11}$$

Example 3.9 You divide 3 pizzas into an equal number of slices each representing $\frac{1}{8}$ of one whole pizza. How many guests can you serve if each guest eats 4 slices?

Solution: First we will find how many total slices are in 3 pizzas by finding $3 \div \frac{1}{8}$.

$$3 \div \frac{1}{8} = \frac{3}{1} \cdot \frac{8}{1} = \frac{24}{1} = 24$$

Then we will find how many guests by dividing 24 slices by 4 for each guest.

$$24 \div 4 = 6$$

You can serve six guests.

CHAPTER 3. MULTIPLYING AND DIVIDING FRACTIONS

☑ Checkpoint - Multiplying and Dividing Fractions

1. Multiply the fractions and reduce to lowest terms.

   a) $\frac{3}{5} \cdot \frac{4}{15}$

   b) $\frac{1}{8} \cdot 12$

   c) $1\frac{1}{2} \cdot 5\frac{4}{7}$

   d) $5 \cdot 3\frac{2}{5}$

   e) $\frac{3}{4} \cdot \frac{8}{9} \cdot \frac{12}{13}$

   f) $\frac{2}{25} \cdot \frac{3}{50} \cdot \frac{4}{75}$

2. Divide the fractions and reduce to lowest terms.

   a) $\frac{3}{5} \div \frac{4}{15}$

   b) $\frac{1}{8} \div 12$

   c) $1\frac{1}{2} \div 5\frac{4}{7}$

   d) $5 \div 3\frac{2}{5}$

   e) $\frac{3}{4} \div \frac{8}{9} \div \frac{12}{13}$

   f) $\frac{2}{25} \div \frac{3}{50} \div \frac{4}{75}$

3. You want to buy $5\frac{1}{2}$ pounds of potatoes for $2/lb. How much will you spend?

4. A butter pecan cookie recipe calls for $1\frac{1}{2}$ cups of pecans for each batch of cookies. How many cups of pecans will you need for 4 batches?

5. You divide 4 pizzas into an equal number of slices each representing $\frac{1}{8}$ of one whole pizza. How many guests can you serve if each guest eats 2 slices?

6. You divide 2 garden plots into rows, dedicating $\frac{1}{6}$ of the plot to each row. Then $\frac{1}{3}$ of the rows will be carrots? How many rows of carrots will you plant?

(Answers in the Appendix.)

Adding and Subtracting Fractions

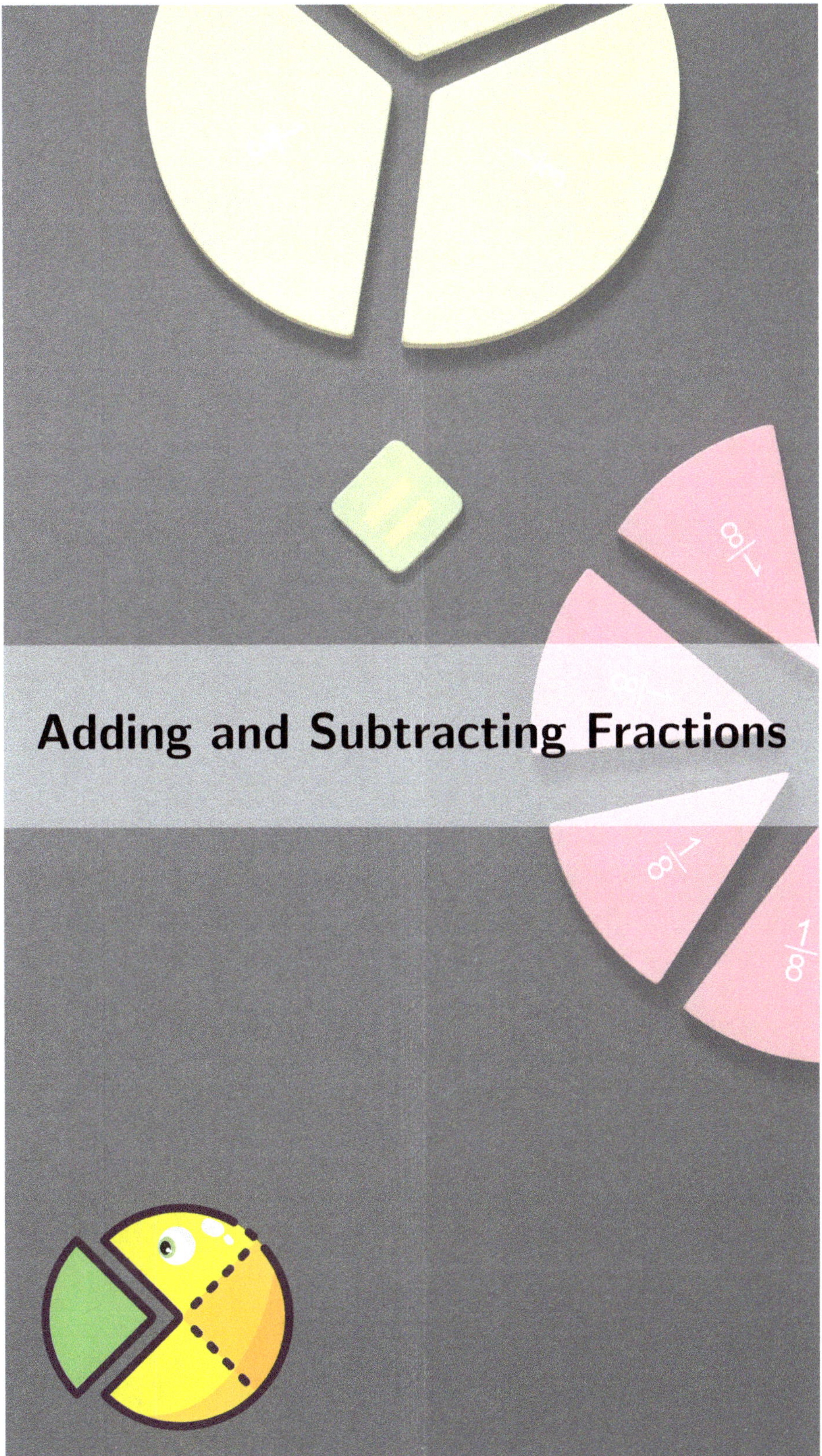

> 🔑 **Key Concepts**
>
> - How to find a lowest common denominator.
>
> - How to add and subtract fractions with a common denominators.
>
> - How to add and subtract fractions with different denominators.

Adding/Subtracting with Common Denominators

When we are **adding or subtracting fractions with common denominators**, we are adding two parts with the *same* whole. For example, when we are adding the following fractions

$$\frac{1}{4} + \frac{2}{4}$$

we are adding the parts 1 and 2 with the same whole of 4. We may illustrate this addition with the following figure and see that the sum is $\frac{3}{4}$.

Therefore, when we are adding fractions 1/4 and 2/4, we are adding the parts and keeping the whole to get 3/4. This process applies, in general, to adding or subtracting fractions with common denominators.

CHAPTER 4. ADDING AND SUBTRACTING FRACTIONS

> ### ⚙ Adding/Subtracting with Common Denominators
>
> Add (or subtract) the numerators and keep the common denominator.
>
> *Example*
> $$\frac{1}{4} + \frac{2}{4} = \frac{1+2}{4} = \frac{3}{4}$$

Example 4.1 Find the sum or difference and reduce to lowest terms.

1. Add $\frac{3}{12} + \frac{9}{12}$
2. Subtract $5\frac{5}{7} - 3\frac{2}{7}$

1. *Solution:* Add the numerators and keep the common denominator
$$\frac{3}{12} + \frac{9}{12} = \frac{3+9}{12} = \frac{12}{12}$$

Reduce to lowest terms.
$$\frac{12 \div 12}{12 \div 12} = \frac{1}{1} = 1$$

2. *Solution:* Change to improper fractions
$$5\frac{5}{7} - 3\frac{2}{7} = \frac{40}{7} - \frac{23}{7}$$

Subtract numerators and keep the common denominator

$$\frac{40-23}{7} = \frac{17}{7}$$

We cannot reduce, since 17 and 7 are both prime numbers, but we can convert the improper fraction back to a mixed number with long division.

$$\frac{17}{7} = 7 \overline{\smash{)}17}^{2} = 2\frac{3}{7}$$
$$\phantom{\frac{17}{7} = 7)}\underline{14}$$
$$\phantom{\frac{17}{7} = 7)00}3$$

> ✏ **In Plain English**
>
> 7 goes into 17, 2 times, with 3 left over.

Example 4.2 You plan to purchase $2\frac{1}{2}$ lbs of grapes and $5\frac{1}{2}$ lbs of potatoes. How many pounds are you purchasing all together?

Solution: We need to find the sum. $2\frac{1}{2} + 5\frac{1}{2}$
Change to improper fractions.

$$2\frac{1}{2} + 5\frac{1}{2} = \frac{5}{2} + \frac{11}{2}$$

Add the numerators and keep the common denominator.

$$\frac{5+11}{2} = \frac{16}{2}$$

Reduce to lowest terms.

53

CHAPTER 4. ADDING AND SUBTRACTING FRACTIONS

$$\frac{16}{2} = 8$$

You will be purchasing 8 pounds all together.

A mixed number is really the sum of a whole number part and a fraction part. For example,

$$2\frac{1}{2} = 2 + \frac{1}{2}$$
$$5\frac{1}{2} = 5 + \frac{1}{2}$$

This means we could also have found the sum from the previous example by adding the whole number parts and the fraction parts separately.

$$2\frac{1}{2} + 5\frac{1}{2} = (2+5) + (\frac{1}{2} + \frac{1}{2}) = 7 + 1 = 8$$

Adding/Subtracting with Different Denominators

When we are **adding or subtracting fractions with different denominators**, we are adding two parts with *different* wholes.

For example,

$$\frac{1}{4} + \frac{1}{2}$$

We are adding different wholes, 4 and 2, with the same part, 1. We may represent this visually with the following illustration of two circles divided into parts:

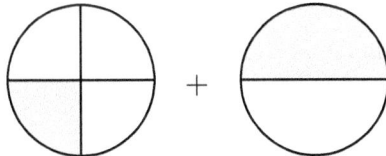

We cannot actually add these fractions as long as they have different wholes. We must first change them to *equivalent fractions* that have the same whole (denominator), and ideally the least common whole/denominator.

A **least common denominator** is the least common multiple of the denominators.

The denominator 4, has multiples **4**, 8, 12, 16, etc.

The denominator 2, has multiples 2, **4**, 6, 8, 10, etc.

The least common multiple of 2 and 4 is 4. Thus, we want both fractions to have least common denominator 4.

We learned earlier that we can *change a fraction to an equivalent fraction by multiplying or dividing both parts by the same amount*. If we multiply both parts of the second fraction by 2 then both fractions will have denominator 4.

$$\frac{1}{4} + \frac{1}{2} \cdot \frac{\mathbf{2}}{\mathbf{2}} = \frac{1}{4} + \frac{2}{4}$$

Visually we would divide the second circle into 4 parts instead of 2 parts, changing our whole from 2 to 4.

CHAPTER 4. ADDING AND SUBTRACTING FRACTIONS

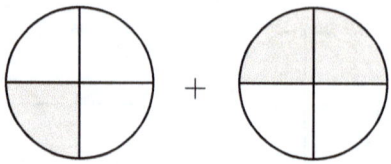

This makes it possible to complete the addition using rules for adding fractions with a common denominator.

$$\frac{1}{4} + \frac{2}{4} = \frac{3}{4}$$

Changing the second circle to 4 parts also makes it easier to see, from the illustration, how the sum is 3 parts out of 4.

In general, we may add or subtract fractions with different denominators using the following rules:

> **Adding/Subtracting with Different Denominators**
>
> - Find the least common denominator (LCD).
>
> - Change to equivalent fractions with the LCD as the denominator.
>
> - Add (or subtract) the numerators and keep the common denominator.

Example 4.3 Find the sum

$$\frac{1}{12} + \frac{1}{18}$$

Solution:

- Find the least common denominator.

  12 has multiples 12, 24, 36, 48, etc.
  18 has multiples 18, 36, 54, 72 etc.
  The least common multiple (denominator) is 36.

- Change to equivalent fractions with 36 as the denominator.

  Multiply the first fraction by 3 to get denominator 36
  Multiply the second fraction by 2 to get denominator 36.

  $$\frac{1}{12} \cdot \frac{\mathbf{3}}{\mathbf{3}} + \frac{1}{18} \cdot \frac{\mathbf{2}}{\mathbf{2}} = \frac{3}{36} + \frac{2}{36}$$

- Add the numerators and keep the common denominator.

  $$\frac{3+2}{36} = \frac{5}{36}$$

We may also **find the least common denominator with prime factorization**.

Find the prime factors of each denominator.

$$\frac{1}{12} + \frac{1}{18} = \frac{1}{2 \cdot 2 \cdot 3} + \frac{1}{2 \cdot 3 \cdot 3}$$

CHAPTER 4. ADDING AND SUBTRACTING FRACTIONS

Both denominators have factors 2·3. Then the first denominator has an additional 2 while the second fraction has an additional 3.

If we multiply the first fraction by 3 and the second fraction by 2

$$\frac{1}{2\cdot 2\cdot 3}\cdot\frac{3}{3}+\frac{1}{2\cdot 3\cdot 3}\cdot\frac{2}{2}$$

we will get common denominator $2\cdot 2\cdot 3\cdot 3$.

$$\frac{1}{2\cdot 2\cdot 3\cdot 3}+\frac{1}{2\cdot 3\cdot 3\cdot 2}$$

Notice that we still end up with common denominator 36 since $2\cdot 2\cdot 3\cdot 3 = 36$.

$$\frac{1}{2\cdot 2\cdot 3\cdot 3}+\frac{1}{2\cdot 3\cdot 3\cdot 2}=\frac{3}{36}+\frac{2}{36}$$

Following are the steps for finding a least common denominator with prime factorization.

> ### ⚙ Finding an LCD with Prime Factorization
>
> - Find the prime factors of each denominator.
> - Multiply the first fraction by the prime factors that are different in the second fraction.
> - Multiply the second fraction by the prime factors that are different in the first fraction.
>
>

Example 4.4 Find the difference. Use prime factorization to find a least common denominator.

$$\frac{1}{16} - \frac{1}{18}$$

- Find the prime factors of each denominator

$$\frac{1}{16} - \frac{1}{18} = \frac{1}{2 \cdot 2 \cdot 2 \cdot 2} - \frac{1}{2 \cdot 3 \cdot 3}$$

- Multiply the first fraction by $3 \cdot 3$, the prime factors that are different in the second fraction.

$$\frac{1}{2 \cdot 2 \cdot 2 \cdot 2} \cdot \frac{\mathbf{3 \cdot 3}}{\mathbf{3 \cdot 3}} - \frac{1}{2 \cdot 3 \cdot 3} = \frac{9}{144} - \frac{1}{2 \cdot 3 \cdot 3}$$

- Multiply the second fraction by $2 \cdot 2 \cdot 2$, the prime factors that are different in the first fraction.

$$\frac{9}{144} - \frac{1}{2 \cdot 3 \cdot 3} \cdot \frac{\mathbf{2 \cdot 2 \cdot 2}}{\mathbf{2 \cdot 2 \cdot 2}} = \frac{9}{144} - \frac{8}{144}$$

- Subtract the numerators and keep the common denominator.

$$\frac{9 - 8}{144} = \frac{1}{144}$$

⚷

Example 4.5 Find the difference.

$$3\frac{1}{9} - 2\frac{1}{12}$$

CHAPTER 4. ADDING AND SUBTRACTING FRACTIONS

- Change to improper fractions

$$3\frac{1}{9} - 2\frac{1}{12} = \frac{28}{9} - \frac{25}{12}$$

- Find the lowest common denominator with prime factorization.

Find the prime factors of each denominator.

$$\frac{28}{9} - \frac{25}{12} = \frac{28}{3 \cdot 3} - \frac{25}{3 \cdot \mathbf{2} \cdot \mathbf{2}}$$

Multiply the first fraction by $2 \cdot 2$ and the second fraction by 3.

$$\frac{28}{3 \cdot 3} \cdot \frac{\mathbf{2 \cdot 2}}{\mathbf{2 \cdot 2}} - \frac{25}{2 \cdot 2 \cdot 3} \cdot \frac{\mathbf{3}}{\mathbf{3}} = \frac{112}{36} - \frac{75}{36}$$

- Subtract the numerators and keep the common denominator.

$$\frac{112 - 75}{36} = \frac{37}{36}$$

- Convert back to a mixed number.

$$\frac{37}{36} = 36 \overline{)37} = 1\frac{1}{36}$$
$$\phantom{\frac{37}{36} = 36\,} \underline{36}$$
$$\phantom{\frac{37}{36} = 36\,\,\,\,} 1$$

> 📝 **In Plain English**
>
> 36 goes into 37, 1 time, with 1 left over.

Example 4.6 A cookie recipe calls for $1\frac{1}{2}$ cups of brown sugar and $\frac{3}{4}$ cups of white sugar. How many total cups are required?

Solution: Find the sum of the cups. $1\frac{1}{2} + \frac{3}{4}$

Change to improper fractions.

$$1\frac{1}{2} + \frac{3}{4} = \frac{3}{2} + \frac{3}{4}$$

Multiply the first fraction by 2 to get lowest common denominator 4.

$$\frac{3}{2} \cdot \frac{2}{2} + \frac{3}{4} = \frac{6}{4} + \frac{3}{4}$$

Add the numerators and keep the common denominator.

$$\frac{6+3}{4} = \frac{9}{4}$$

Convert to a mixed number.

$$\frac{9}{4} = 4\overline{)9} = 2\frac{1}{4}$$

The recipe requires $2\frac{1}{4}$ cups of sugar.

CHAPTER 4. ADDING AND SUBTRACTING FRACTIONS

☑ Checkpoint - Adding and Subtracting Fractions

1. Find each sum or difference.

   a)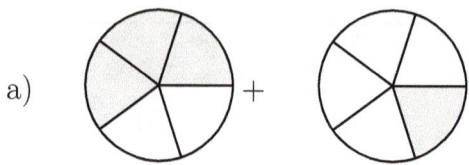

   b)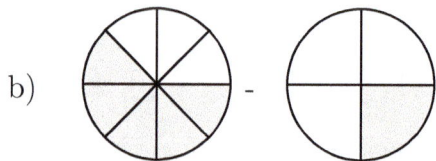

2. Find each sum or difference.

   a) $\frac{3}{5} + \frac{1}{5}$

   b) $\frac{1}{8} + 12$

   c) $1\frac{1}{2} + 5\frac{4}{7}$

   d) $5 - 3\frac{2}{5}$

   e) $\frac{37}{100} - \frac{12}{100}$

   f) $\frac{2}{25} - \frac{4}{75}$

3. You are making two batches of butter pecan cookies and each batch requires $2\frac{2}{3}$ cups of flour. How many cups of flour will you need for two batches?

4. You have one 40 pound bag of top soil that contains $\frac{3}{4}$ cubic feet of soil. If you need $16\frac{1}{2}$ cubic feet of soil for a garden plot, how many more cubic feet of soil do you need? How many 40 pound bags should you buy?

(Answers in the Appendix.)

> ⚠ **CAUTION**
>
> Do not go beyond this point until you are comfortable with adding, subtracting, multiplying and dividing fractions without variables.

Fractions Containing Variables

> **🔑 Key Concepts**
> - Reducing Fractions with Variables
> - Multiplying Fractions with Variables
> - Dividing Fractions with Variables
> - Adding and Subtracting Fractions with Variables

Introduction to Fractions with Variables

A **variable** (🔑) is a letter (e.g. x, y, m, n, etc.) that holds the place of a number. It is called a variable because its value may vary based on the number it represents.

Consider the expression, $\frac{1}{2}x$. In English, this expression means "$\frac{1}{2}$ times x". If we know what value x represents, we can multiply that value by a $\frac{1}{2}$ to find $\frac{1}{2}x$.

> If x = 4, then $\frac{1}{2}x = 2$, because "$\frac{1}{2}$ times 4" is 2.
>
> If x = 6, then $\frac{1}{2}x = 3$, because "$\frac{1}{2}$ times 6" is 3.
>
> If x = 8, then $\frac{1}{2}x = 4$, because "$\frac{1}{2}$ times 8" is 4.

Notice that, because of x, the value of the expression $\frac{1}{2}x$ may vary as x varies.

CHAPTER 5. FRACTIONS WITH VARIABLES

> 💡 **Math Bit**
>
> The variable in a fraction term[a] ($\hookleftarrow$) may be placed in the numerator of the fraction or on the right side of the fraction.
>
> $$\frac{1}{2}x = \frac{x}{2}$$
>
> ---
> [a] A number, a variable, or a number times some variable(s). (e.g. 3, 5x, 9xy, etc.)

Variables are important because they help us make general calculations for different contexts, and we use them in many different situations. Even though we may not always know it. For example, if we want to make something with a recipe, but only use half of all the ingredients we will be using the expression $\frac{1}{2}x$ to find the correct amount for each ingredient.

> If x = 2 cups of flour, then $\frac{1}{2}x$ = 1 cup of flour
>
> If x = 6 quarts of oil, then $\frac{1}{2}x$ = 3 quarts of oil
>
> If x = 4 lbs. of chicken then $\frac{1}{2}x$ = 2 lbs. of chicken.

Variables give us a way to communicate general calculations in different situations.

Reducing Fractions with Variables

We can reduce a fraction containing variables if both parts share common number or variable factors.

> ⚙ **Reducing Fractions with Variables**
>
> To reduce a fraction with variables to lowest terms
>
> - Write each part of the fraction as a product of its prime factors and variables.
>
> - Cancel common factors and variables.

Example 5.1 Reduce each fraction to lowest terms.

1. $\dfrac{6xy}{8x}$
2. $\dfrac{14mn}{21n}$

1. *Solution:* Write each part of the fraction as a product of its prime factors and its variables.

$$\frac{6xy}{8x} = \frac{2 \cdot 3xy}{2 \cdot 2 \cdot 2x}$$

Cancel common factors and variables.

$$\frac{\cancel{2} \cdot 3\cancel{x}y}{\cancel{2} \cdot 2 \cdot 2\cancel{x}} = \frac{3y}{4}$$

2. *Solution:* Write each part of the fraction as a product of its prime factors and its variables.

$$\frac{14mn}{21n} = \frac{2 \cdot 7mn}{3 \cdot 7n}$$

Cancel common factors and variables.

$$\frac{2 \cdot \cancel{7}m\cancel{n}}{3 \cdot \cancel{7}\cancel{n}} = \frac{2m}{3}$$

CHAPTER 5. FRACTIONS WITH VARIABLES

Adding/Subtracting Fractions with Variables

To add or subtract fractions containing variables, we must know how to combine variable terms.

> ⚙ **Combining Variable Terms**
>
> Combine the number parts and *keep* the common variable part.

Example 5.2 Simplify each expression by combing the variable terms.

1. $2x + 3x$
2. $12xy - 9xy$

1. *Solution:* Combine the number parts and keep the common variable part.

$$2x + 3x = (2+3)x = 5x$$

2. *Solution:* Combine the number parts and keep the common variable part.

$$12xy - 9xy = (12-9)xy = 3xy$$

> ⚠ **CAUTION**
>
> We cannot combine terms that do not have the same variable part.
>
> $$5x + 7y \neq 12xy$$
>
> $$8 + x \neq 8x$$

We can add or subtract fractions containing variables with the following process:

> **⚙ Adding or Subtracting Fractions with Variables**
>
> - Move each variable on the right to the numerator of its corresponding fraction.
>
> - If necessary, change to equivalent fractions with a least common denominator.
>
> - Combine the numerators and keep the common denominator.
>
> - Reduce to lowest terms.

Example 5.3 Add the fractions containing variables.

1. $\frac{1}{8}x + \frac{4}{8}x$
2. $\frac{1}{2}xy + \frac{1}{4}xy$

1. *Solution:* Move each variable on the right to the numerator of its corresponding fraction.

CHAPTER 5. FRACTIONS WITH VARIABLES

$$\frac{1}{8}x + \frac{4}{8}x = \frac{1x}{8} + \frac{4x}{8}$$

Combine the numerators and keep the common denominator.

$$\frac{1x + 4x}{8} = \frac{5x}{8} = \frac{5}{8}x$$

2. *Solution:* Move each variable on the right to the numerator of its corresponding fraction.

$$\frac{1}{2}xy + \frac{1}{4}xy = \frac{1xy}{2} + \frac{1xy}{4}$$

Change to equivalent fractions with least common denominator 4.

$$\frac{1xy}{2} \cdot \frac{2}{2} + \frac{1xy}{4} = \frac{2xy}{4} + \frac{1xy}{4}$$

Combine the numerators and keep the common denominator.

$$\frac{2xy}{4} + \frac{1xy}{4} = \frac{2xy + 1xy}{4} = \frac{3xy}{4}$$

It is also possible to add the fractions in the previous example by combining the number parts and keeping the common variable part.

$$\frac{1}{8}x + \frac{4}{8}x = (\frac{1}{8} + \frac{4}{8})x = \frac{5}{8}x$$

$$\frac{1}{2}xy + \frac{1}{4}xy = (\frac{1}{2} + \frac{1}{4})xy = \frac{3}{4}xy$$

Example 5.4 Add the fractions containing variables.

1. $\frac{1}{8} + \frac{1}{x}$
2. $\frac{1}{x} + \frac{1}{y}$

1. *Solution:* Change to equivalent fractions with lowest common denominator.

    Multiply the 1st fraction by the factor that is missing from the 2nd denominator; Multiply the 2nd fraction by the factor that is missing from the 1st denominator.

    $$\frac{1}{8} \cdot \frac{x}{x} + \frac{1}{x} \cdot \frac{8}{8} = \frac{x}{8x} + \frac{8}{8x}$$

    Combine the numerators and keep the common denominator.

    $$\frac{x}{8x} + \frac{8}{8x} = \frac{x+8}{8x}$$

    We leave the numerator as "x + 8" because we cannot combine terms that do not have the same variable part.

2. *Solution:* Change to equivalent fractions with lowest common denominator.

    Multiply the 1st fraction by the factor that is missing from the 2nd denominator; Multiply the 2nd fraction by the factor that is missing from the 1st denominator.

CHAPTER 5. FRACTIONS WITH VARIABLES

$$\frac{1}{x} \cdot \frac{y}{y} + \frac{1}{y} \cdot \frac{x}{x} = \frac{y}{xy} + \frac{x}{xy}$$

Combine the numerators and keep the common denominator.

$$\frac{y}{xy} + \frac{x}{xy} = \frac{y+x}{xy}$$

Multiplying and Dividing Fractions with Variables

When we combine variable terms, the number in front tells us how many we have of what the variable represents.

> x = 1x
>
> x + x = 2x
>
> x + x + x = 3x
>
> x + x + x + x = 4x
>
> etc.

When we multiply variables we use an exponent to indicate *how many factors* we have of what the variable represents.

> $x = x^1$
>
> $x \cdot x = x^2$

$$x \cdot x \cdot x = x^3$$

$$x \cdot x \cdot x \cdot x = x^4$$

etc.

A term like, x^2, is what we call an exponent expression where x is called the base and 2 is called the **exponent** (✎).

We use exponents to multiply fractions containing variables.

> **Multiplying Fractions with Variables**
>
> - Move each variable on the right to the numerator of its corresponding fraction.
> - Multiply across
> - Reduce to lowest terms.

Example 5.5 Multiply the fractions containing variables.

1. $\dfrac{x}{2} \cdot \dfrac{x}{8}$
2. $\dfrac{1}{5}m^2 \cdot \dfrac{2}{m}$

1. *Solution:* Multiply across

$$\frac{x}{2} \cdot \frac{x}{8} = \frac{x \cdot x}{2 \cdot 8} = \frac{x^2}{16}$$

2. *Solution:* Move each variable on the right to the numerator of its corresponding fraction.

CHAPTER 5. FRACTIONS WITH VARIABLES

$$\frac{1}{5}m^2 \cdot \frac{2}{m} = \frac{1m^2}{5} \cdot \frac{2}{m}$$

Multiply across.

$$\frac{1m^2}{5} \cdot \frac{2}{m} = \frac{1 \cdot 2 \cdot m^2}{5 \cdot m} = \frac{2m^2}{5m}$$

Reduce to lowest terms by canceling common factors.

$$\frac{2m^2}{5m} = \frac{2m \cdot \cancel{m}}{5\cancel{m}} = \frac{2m}{5}$$

It is also possible to cancel common factors before multiplying.

$$\frac{1m^2}{5} \cdot \frac{2}{m} = \frac{1m \cdot \cancel{m}}{5} \cdot \frac{2}{\cancel{m}} = \frac{1m}{5} \cdot \frac{2}{1}$$

Then multiply across.

$$\frac{1m}{5} \cdot \frac{2}{1} = \frac{2m}{5}$$

When we divide fractions containing variables we change the division problem to a multiplication problem and apply the sames steps we used for multiplying fractions with variables.

⚙ Dividing Fractions with Variables

- Move each variable on the right to the numerator of its corresponding fraction.
- Flip the 2nd fraction and change to multiplication.
- Multiply across
- Reduce to lowest terms.

Example 5.6 Divide the fractions containing variables.

1. $\frac{1}{2}x \div \frac{4}{7}x$
2. $\frac{12}{y} \div \frac{12}{x}$

1. *Solution:* Move each variable on the right to the numerator of its corresponding fraction.

$$\frac{1}{2}x \div \frac{4}{7}x = \frac{x}{2} \div \frac{4x}{7}$$

Flip the 2nd fractions and change to multiplication.

$$\frac{x}{2} \div \frac{4x}{7} = \frac{x}{2} \cdot \frac{7}{4x}$$

Multiply across

$$\frac{x}{2} \cdot \frac{7}{4x} = \frac{7x}{8x}$$

Reduce to lowest terms by canceling common factors.

$$\frac{7\cancel{x}}{8\cancel{x}} = \frac{7}{8}$$

CHAPTER 5. FRACTIONS WITH VARIABLES

2. *Solution:* Flip the 2nd fraction and change to multiplication.

$$\frac{12}{y} \div \frac{12}{x} = \frac{12}{y} \cdot \frac{x}{12}$$

Multiply across.

$$\frac{12}{y} \cdot \frac{x}{12} = \frac{12x}{12y}$$

Reduce to lowest terms by canceling common factors.

$$\frac{\cancel{12}x}{\cancel{12}y} = \frac{x}{y}$$

☑ Checkpoint - Fractions Containing Variables

1. Fill in the blank with a **T** if the statement is *true* and an **F** if the statement is *false*.

   ___ (a) $3 + x = 3x$ ___ (e) $3x = x + x + x$

   ___ (b) $x = 1x$ ___ (f) $x^5 = x \cdot 5$

   ___ (c) $x^5 = x \cdot x \cdot x \cdot x \cdot x$ ___ (g) $x^2 \cdot x^3 = x^6$

   ___ (d) $x^2 \cdot x^3 = x^5$ ___ (h) $\frac{1}{5}x = \frac{x}{5}$

2. Reduce each fraction to lowest terms.

a) $\frac{5x}{10x}$ c) $\frac{18m}{36mn}$

b) $\frac{27xy}{54y}$ d) $\frac{n}{mn^2}$

3. Combine the fractions containing variables and reduce to lowest terms.

a) $\frac{x}{5} + \frac{x}{5}$ d) $1\frac{1}{3}n - \frac{n}{3}$

b) $\frac{m}{4} - \frac{m}{8}$ e) $\frac{10}{x} - \frac{9}{x}$

c) $\frac{7}{8} + \frac{4}{x}$ f) $\frac{1}{m} - \frac{1}{n}$

4. Multiply the fractions containing variables and reduce to lowest terms.

a) $\frac{x}{5} \cdot \frac{x}{5}$ d) $1\frac{1}{3}n \cdot \frac{n}{3}$

b) $\frac{m}{4} \cdot \frac{m}{8}$ e) $\frac{10}{x} \cdot \frac{9}{x}$

c) $\frac{7}{8} \cdot \frac{4}{x}$ f) $\frac{1}{m} \cdot \frac{1}{n}$

5. Divide the fractions containing variables and reduce to lowest terms.

a) $\frac{x}{5} \div \frac{x}{5}$ d) $1\frac{1}{3}n \div \frac{n}{3}$

b) $\frac{m}{4} \div \frac{m}{8}$ e) $\frac{10}{x} \div \frac{9}{x}$

c) $\frac{7}{8} \div \frac{4}{x}$ f) $\frac{1}{m} \div \frac{1}{n}$

(Answers in the Appendix.)

Appendix

Checkpoint Answers

Introduction to Fractions

1. a) $<$
   b) $<$
   c) $<$
   d) $=$
   e) $>$
   f) $<$

2. a) $\frac{38}{7}$
   b) $\frac{15}{4}$
   c) $3\frac{1}{7}$
   d) $8\frac{2}{5}$

3. a) $\frac{5}{25}$
   b) $\frac{1}{25}$
   c) $\frac{12}{25}$
   d) $\frac{13}{25}$

4. a) $\frac{3}{10}$
   b) $\frac{7}{10}$

Reducing Fractions

1. a) 8
   b) 5
   c) 2
   d) 14

2. a) $2 \cdot 2 \cdot 2 \cdot 2 \cdot 2$
   b) $2 \cdot 2 \cdot 3 \cdot 7$
   c) $3 \cdot 3 \cdot 3 \cdot 5$
   d) $5 \cdot 5 \cdot 5 \cdot 5$

3. a) $\frac{1}{3}$
   b) $\frac{4}{7}$
   c) $\frac{27}{125}$
   d) $\frac{8}{21}$

Multiplying and Dividing Fractions

1. a) $\frac{4}{25}$

   b) $\frac{3}{2}$ or $1\frac{1}{2}$

   c) $\frac{117}{14}$ or $8\frac{5}{14}$

   d) 17

   e) $\frac{8}{13}$

   f) $\frac{4}{15625}$

2. a) $\frac{9}{4}$ or $2\frac{1}{4}$

   b) $\frac{1}{96}$

   c) $\frac{7}{26}$

   d) $\frac{25}{17}$ or $1\frac{8}{17}$

   e) $\frac{117}{128}$

   f) 25

3. $11

4. 6 cups

5. 16 guests

6. 4 rows of carrots

Adding and Subtracting Fractions

1. a) The sum is illustrated below:

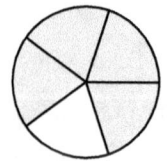

b) The difference is illustrated below:

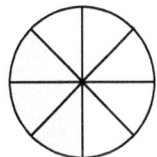

2. a) $\frac{4}{5}$　　　　d) $\frac{8}{5}$ or $1\frac{3}{5}$

　b) $\frac{97}{8}$ or $96\frac{1}{8}$　　e) $\frac{1}{4}$

　c) $\frac{97}{8}$ or $12\frac{1}{8}$　　f) $\frac{2}{75}$

3. $5\frac{1}{3}$ cups of flour

4. $15\frac{3}{4}$ cubic feet more; 21 bags

Fractions Containing Variables

1. a) F　　　　e) T
　b) T　　　　f) F
　c) T　　　　g) F
　d) T　　　　h) T

2. a) $\frac{1}{2}$　　　　c) $\frac{2}{n}$
　b) $\frac{x}{2}$　　　　d) $\frac{1}{mn}$

3. a) $\frac{2x}{5}$　　　　d) n

　b) $\frac{m}{8}$　　　　e) $\frac{1}{x}$

　c) $\frac{7x+32}{8x}$　　　f) $\frac{n-m}{mn}$

4. a) $\frac{x^2}{25}$　　　　d) $\frac{4n^2}{9}$

　b) $\frac{m^2}{32}$　　　　e) $\frac{90}{x^2}$

　c) $\frac{7}{2x}$　　　　f) $\frac{1}{mn}$

5. a) 1　　　　d) 4

　b) 2　　　　e) $\frac{10}{9}$ or $1\frac{1}{9}$

　c) $\frac{7x}{32}$　　　f) $\frac{n}{m}$

Study Questions

Instructions: Complete these study questions to get a better understanding of key fraction concepts. Answers can be found in the text. Search for the 🔍 symbol to find the information you need.

Comparing Fractions

1. Fill in the tables with an example/definition for each:

Fraction	Proper Fraction

Improper Fraction	Mixed Number

2. Fill in the blanks.

   To change an **improper fraction to a mixed number** we must divide the _____ by the _____.

   > *Example:*

   To change a **mixed number to an improper fraction** we must multiply the _____ and the _____. Then add the _____, and place the result over the original denominator.

   > *Example:*

3. Two fractions are _____ if they represent the same numerical value.

   > *Example:*

Reducing Fractions

1. Fill in the tables with an example/definition for each:

Greatest Common Divisor

Prime Factor

2. **To reduce with a greatest common divisor**, divide the numerator and the denominator by the _____.

   $\frac{24}{32} =$

   Answer $= \frac{3}{4}$

3. **To reduce with prime factorization**, write the numerator and the denominator as a product of its _____ factors, and _____ all common factors between the numerator and the denominator.

$$\frac{24}{32} =$$

Answer = $\frac{3}{4}$

Operations of Fractions

1. Fill in the blanks and show work for the example given under each operation.

    When **multiplying fractions** multiply _____ and reduce to lowest terms.

    $$\frac{3}{4} \cdot \frac{2}{9} =$$

    Hint: See Example 3.1

    When **dividing fractions** _____ the 2nd fraction and multiply.

$$\frac{3}{4} \div \frac{9}{12} =$$

Hint:See Example 3.6

When **adding/subtracting fractions with common denominators** combine the numerators and _____ the common denominator.

$$\frac{3}{12} + \frac{9}{12} =$$

Hint:See Example 4.1

When **adding/subtracting fractions with different denominators**, change to equivalent fractions with a _____ and add/subtract.

$$\frac{1}{12} + \frac{1}{18} =$$

Hint:See Example 4.3

2. Give the steps for **finding an LCD with Prime Factorization**. (Hint: See Example 4.4)

3. Find the LCD with prime factorization and subtract, showing work in the box below.

$$\frac{1}{14} - \frac{1}{21} =$$

Fractions with Variables

1. Fill in the table with an example/definition for each.

Variable	Term

Exponent

2. When **combining variable terms** _____ the number parts and _____ the common variable part.

2x + 3x =
12xy - 9xy =

3. Show how to get each answer.

$$\frac{m}{4} - \frac{m}{8} =$$

Answer: $\frac{m}{8}$

$$\frac{10}{x} \div \frac{9}{x} =$$

Answer: $1\frac{1}{9}$

Index

composite numbers, 28

denominator, 6
divisibility rules, 24

exponent, 73

factor, 28
fractions, 7

greatest common divisor, 26

improper fraction, 14

mixed fraction, 15
mixed number, 15

numerator, 6

prime factor, 28
prime factorization, 28
prime numbers, 28
proper fraction, 14

term, 66

variable, 65

About the Author

Hello! My name is Eve Wallis. I am an award-winning teacher with a master's degree in math and more than a decade of experience teaching adult learners. The courses I have taught include high school and college algebra, Calculus, technical math, math in society, and student success. I enjoy writing and reading books, learning new things, and living in the mountains with my husband and our two children.

I hope you enjoyed this first book on Fractions. To learn more about the books I am writing visit my website at

HTTPS://EVEWALLIS.WORDPRESS.COM

Message of Hope

So you have made a decision to learn something about math. I am so glad! Learning math is important, because math is an essential part of your daily life.

Another essential part of life is your decision about what comes next. If you haven't already, I hope you will consider God's grace provision made available to you, in this life, to provide for your future. You know your time on this earth is temporary, but did you know that you have the opportunity for an everlasting life in heaven? All you must do is make one simple decision to believe in the salvation of Jesus Christ.

The Bible tells us in John 3:16 NET, "For this is the way God loved the world: He gave his one and only Son, so that everyone who believes in him will not perish but have eternal life. "

You can make a decision to believe in Jesus Christ right here, right now, knowing that God's grace gift of salvation through His Son is sufficient to pay the penalty of sin and open the door to a relationship with God and a life everlasting.

"I am the door. If anyone enters through me, he will be saved, and will come in and out and find pasture."
-John 10:9 NET

www.ingramcontent.com/pod-product-compliance
Lightning Source LLC
LaVergne TN
LVHW021600070426
835507LV00014B/1877